PROPOLIS

by
Klaus Nowottnick

Northern Bee Books

Propolis

© Klaus Nowottnick

Many of the recipes referred to in this volume have their origins in long practised folk medicines from Eastern Europe.

However readers should note that the publishers have no professional knowledge that any of the claims made for the powers ascribed to propolis are correct and they are strongly advised to take medical advice before using any of the enclosed recipes to dose themselves.

ISBN 978-1-908904-15-7

PUBLISHED BY
Northern Bee Books
Scout Bottom Farm
Mytholmroyd
Hebden Bridge HX7 5JS (UK)
www.GroovyCart.co.uk/beebooks

PROPOLIS

by
Klaus Nowottnick

In gratitude and love for my dear girl friend
Mrs Katrin Flach from Mittelsdorf

TABLE OF CONTENTS

TABLE OF CONTENTS

TABLE OF CONTENTS

PREFACE

Nearly 20 years have passed since the first edition of this book was published. Originally in German, published by Leopold-Stocker–Verlag in Graz, Austria in 1987, I never expected much interest in my little book. But very soon the first edition was sold out: indeed it had found widespread acceptance. In 1996 the Czech and Slovak editions were published and in the summer of 2003 the 3rd German edition came out.

The first edition was based predominantly on the research being done in the Eastern European countries. Since then, it turns out that extensive research has taken place in Western Europe: especially in France and Denmark. Also some German researchers dedicated themselves to the study of propolis, although they are few, most notably Professor Havsteen of the University of Kiel, but even this limited research has given us satisfaction and confidence.

I have expanded the number of formulas and recipes, hopefully a good thing, and left to each reader the opportunity to personalize and customize them. Always get the advice and consent of a doctor where serious complaints exist because marketing propolis products for medical use is only allowed in a few countries.

Klaus Nowottnick

PROPOLIS THROUGHOUT HISTORY

Consisting of the Greek word Pro, which means "before" and Polis which means "city," lie the origins of the word propolis; literally representing that which the bees produce in front of their hive. Colloquially this bee product is very often referred to as "bee glue." In 1910, Heyses "Dictionary of borrowed words" published by the German company Hahnsche, Verlagsbuchhandlung, Hannover and Leipzig, propolis was defined as: A suburb, a porch of the bee hive.

Propolis has been known for thousands of years. Marcus Terentius Varro (116-27 B.C.) wrote:

> *They call that 'Propolis' out of which the bees make a protection in the hive's entrance in high summer. It is used with the same name by doctors for poultices, for which reason it is more expensive than honey in the Via Sacra.*

Other ancient writers studied bee products and knew their uses including Pliny (23-79 A.D.) and Dioscorides (40-90 A.D.) They both referred to this adhesive material as propolis. The Greek philosopher Aristotle (384-322 B.C.) wrote of his knowledge of propolis when he built a hive with transparent walls. He wanted to observe and study the life of the bee colony in detail, hoping to make conclusions about the social community of humans. However, the bees did not want to reveal their mysteries and covered the transparent walls of the hive with a shellac-like substance, most likely propolis ("The Holy Bee," G. Ransom).

In his medical works, Ali Ibn Sina (Avicenna 930–1037 A.D.) distinguishes two kinds of wax, one as "pure wax", which forms honeycomb, in which the bees take care of their brood and store honey; while the "black wax" is marked as refuse of the hive. Along with "black wax," propolis was clearly thought to be in this category. Avicenna confesses, "The characteristic of this stuff is, to pull out the tips of the arrows and the thorns, it refines and cleans the skin and softens it."

In the Balkan countries, the wood carvers still mix Propolis with grain alcohol for painting their carvings because this mixture creates an especially luminous yellow.

It is true Italian violin-makers could increase their instrument's tension and firmness by propolis varnish, helping them create their warm and melodious sound. The world-famous Stradivarius violins got their dark color by the addition of "dragon blood," the reddish resin from the fruit of the Dragon Blood palm tree, added to propolis varnish.

Makashvili determined that propolis was known by the priests of ancient Egypt who used it in the chemistry and art of mummification. In Grusinian medicinal essays of the 12th – 15th century, the healing power of propolis is mentioned frequently. Here is an example from the Grusinian medicine book "Karabadine," by Zaza Fanaskerteli-Tizischwilli, where he proposes a remedy against inflammations of the mouth and for treating cavities:

> *One takes propolis, adds a little arsenic, red lentil,*
> *yarrow and germander and pulverizes and strains*
> *it. Then a spoonful of olive oil and honey is added.*
> *Everything is mixed well and placed on the ill tooth.*

In Arabia propolis was used for the treatment of toothaches, cavities and mouth inflammations. The propolis was finely ground and mixed with olive oil and rubbed on the affected area. It has been told this mixture has an analgesic and anti-inflammatory effect. Napoleon's troops are reported to have used propolis for treatment of injuries.

Grusinian folk medicine used ointments made with propolis to combat diseases. New-born children had a propolis cake applied to the umbilicus. People long ago coated children's wooden toys with propolis, keeping them sterile, which under the un-hygienic conditions of the time had a special importance. The popularity of propolis is understandable because it was recognized by "doctors" and because beekeepers advocated and used it. Every beekeeper knows that the Caucasian honey bee is the best "Propolizer," important to know for both the producer and consumer of propolis.

Specialists who have researched Grusinian medicine have come to the conclusion that the knowledge carried from generation to generation has a value and their folk medicine should be preserved. Unique recipes were treated as family secrets and transmitted from one generation to another.

Popular Grusinian recipes:

i.   In case of pain caused by a cold, a warmed propolis cake is helpful. It should be placed on the painful area.
ii.   Rheumatism pain in the extremities can be eliminated by applying a warmed propolis cake over night.
iii. For a boil, place a thin, warmed propolis sheet over the inflamed area. After a short time, remove the poltice and the infected liquid below.
iv.   For corns, feet should be soaked in warm water, and then thin, warm sheets of propolis should be applied and bandaged. The corn, with its root, will fall out in a short while.

Most beekeepers do not want their bees to be heavy propolizers, since it causes more work and takes more time to work the hive. The management of colonies becomes a true torment for the beekeeper and the bees. Different bee races show different propolizing behavior. In Germany, Carniolans have been predominately used because they produce less propolis. However, beekeepers have developed an increased interest in propolis due to the discoveries of beneficial characteristics and propolis is now being collected for personal as well as commercial use. Warren Ogren wrote that there is more money to be made from selling hive scrapings for the propolis they contain than rendering down the wax, which leaves the propolis useless. His business purchases thousands of dollars worth of cappings, hive scrapings and burr comb annually to extract the propolis. From 100 pounds of scrapings, 50-60 pounds is pure beeswax. But if the beekeeper takes the time and effort to sort out the propolis from the wax, he will make more money from the two commodities separately than he would selling just the rendered wax.

There is an increasing demand for natural and pure foods due to a growing aversion to industrially produced products containing artificial fragrances, flavor enhancers and preservatives. There are increasing numbers of allergies as a result of these synthetic additives. As a result, there is an immense diversity of new allergy medicines appearing on the market. Their effectiveness in many cases is impressive, but the damage caused by long-term usage will be deemed hazardous. The effect of using naturally produced substances is completely different. Of course, the curative effects do not manifest themselves as rapidly, but the success is considerable and side-effects almost unnoticeable. Propolis is one of these natural remedies and it is one of the most effective. Slowly but surely, the study and use of apitherapy is increasing on a global scale. Some years ago, the German Apitherapy Association was founded, which has members from Germany, many European countries and even America. The bees do not produce propolis for the beekeeper, but for themselves. Propolis is what guarantees the survival of the bee colony. Because of the antibacterial characteristics of propolis, the bee colony is able to survive. In the beehive, propolis is used for smoothing the rough parts of the hive walls and for the sealing and closing of leaky places. In the tropical and subtropical regions of our earth, the bees use Propolis to erect barriers and deflections within the entrance, to protect themselves from enemies.

We have found no confirmation supporting the commonly held belief that the bee's tendency to propolize is a defense action against cold and wind. The propolizing behavior of the bees probably serves the same purpose in moderate climates as in warm climates. If the bees kill an intruder too large to be removed from the hive, they will propolize the body. Mummified animals, like mice, lizards and larger butterflies, are hermetically sealed, and the development and growth of dangerous germs is contained or inhibited. Brother Adam, during his extensive journeys in search of the best strains of bees, discovered that the bee races of Asia have a distinct tendency to propolize.

COLLECTING STRATEGIES OF THE HONEY BEES   2.1

Primarily, it is only a few older bees of the colony which do the difficult job of collecting this resinous and adhesive material. These bees have the specialized task of collecting resins on warm days between 10:00 am and 4:00 pm. When working a productive nectar flow, the colony won't devote much attention to propolis. This is important for the comb honey producer to take into consideration; during the time of peak nectar flow, the honey combs will be propolized less and therefore the fresh wax comb will have a pure white color.

We know that propolis is produced by two methods. For as long as we have known, bees have collected resin from the buds and branches of the poplar and willow, the buds of birch, alder and sweet chestnut, some herbs and to a smaller extent, conifer buds (pine and spruce.)

It is also believed that a large portion of propolis is produced as a by-product of pollen digestion. The pollen cover, or exine, contains materials and oils which protect the inner substance of the pollen grain. When the bees consume the pollen to prepare larval food, they have to open the pollen grain by removing the hard and indigestible cover of the pollen grain, which is excreted as propolis droplets.

The bees who are producing propolis rarely engage in other activities. They are mostly older than 15 days and they can make only a few flights daily because of the strenuousness of their work. A resin collecting flight normally lasts for 15 - 20 minutes. After they return to the hive, the propolis collector takes a longer break and consumes more honey that her nectar and pollen collecting counterparts. Then she starts again on a new collecting flight. In contrast to the haste of the nectar and pollen collecting bees, the propolis bee goes quietly about her work. Apparently the resinous material does not pose as much of a problem to the bees as we might think because they do not compress the resinous material and they do not radiate body heat. Nevertheless, gathering this natural glue is still monumental work. There are very impressive step by step photos, which illustrate the efforts and troubles of the bees during collection. Also, they display the bees' deftness of touch when handling this gooey material.

The bee stretches the resin with all her might, like a rubber band. She grips the base with the middle and hind legs, with the thorax set high and the tip of the abdomen touching the bottom. With her head stretched, one can see the white, otherwise not visible, sinew band of the throat. The antennas circle like the rotors on a helicopter over the active mandibles.

Fig 1 The bees will fill scratches and crevices with propolis, unfortunately the mobility of the hive parts is impaired and complicates the work of the beekeeper. (Photo Nowottnick)

Fig 2 The bees have constructed a barrier of propolis in the hive entrance. (Photo Nowottnick)

Fig 3   A propolis collecting honey bee at a frame with propolis. (Photo Nowottnick).

Fig 4, 5, 6 A propolis collecting bee on a poplar tree. (Photos Syx from "In the empire of
the bees", Ehrenwirth - publishing house Munich).

Using the antenna, the bee discovers a particle of her objective, and will use the mandibles to stretch it and then the work begins.

The resin bit will be stretched until it eventually breaks. Cleverly, and yet with tremendous effort, the adhesive substance will then be transported to the hind legs and placed in the pollen baskets.

When the collector bee returns to her home hive, quite often she will be greeted at the entrance and relieved of her load. But as a rule, the bee herself transports the propolis to the place in the hive where it will be needed and there it is removed and processed by her sisters.

Propolis becomes very flexible at a temperature of 68°F (20°C), or higher, and melts at approx. 212°F (100°C).

According to Russian bee scientists, each pollen basket holds a propolis load of approximately 10mg (0.00035274 oz.) The amount collected differs greatly depending on the climatic and geographical situation of the apiary as well as the available vegetation. Not to be overlooked are the variances between the strains of bees.

COMPOSITION OF PROPOLIS   2.2

Because of the differing regional sources, propolis does not have a constant chemical composition and therefore it is impossible to bind it to a chemical formula. Propolis consists of approximately 55% resin and balm materials, about 30% wax, 5-10% ethereal oils, about 2-5% pollen and the remaining parts are vitamins and micro-elements. Propolis balm contains cinnamic alcohol, cinnamic acid and tannins. It has been proved to contain vitamin B1 (Thiamine, Niacin), provitamin A and a series of other trace elements. The colors are different depending on the origin, ranging between bright brown to dark red. Propolis keeps it's efficacy up to five years when stored in airtight packaging in a cool, dark place. But propolis is most effective when fresh.

Propolis produced in moderate climates possesses about 40 different phenol components, of which about 90-95% are flavonoids, which are known for their immune strengthening characteristics. The primary benefits of propolis are from its broad-spectrum anti-bacterial, anti-viral and anti-fungal properties. Unpleasant side-effects are very rare, as opposed to traditional antibiotics. The Ferula acid in propolis creates the antibacterial effect (CIZMARIK u. MANTEL). The appearance of such vital trace elements like iron, aluminium, vanadium, calcium, silicon, manganese, strontium, sodium and magnesium is essential for an optimal performance of the physiological processes. Their sphere of influence extends from the metabolism of protein, fat and sugar, the body's ability to synthesize protein, the thermal modulation of the body, on blood and bone build up and on the body's biological immune reactions.

The following diagram compares testing of the affects of various plant parts of the black cottonwood (*Populus nigra* L.) and propolis on *Bacilli subtilis*.

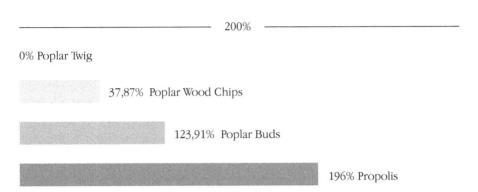

———————————————————— 200% ————————————————

0% Poplar Twig

37,87% Poplar Wood Chips

123,91% Poplar Buds

196% Propolis

Fig 7   The antibiotic value of propolis in comparison with parts of the poplar tree.

Research conducted in France produced some surprising results; when testing various insects for known, parastitic bacteria, (CHAUVIN), they determined that bees were absolutely parasite free. Neither bacteria nor viruses can live their parasite existence on the bees' body, because the bees secrete an antibiotic. Further tests demonstrated the antimicrobic effect of propolis on many micro-organisms. Some bacterial strains are destroyed within 15-20 minutes, others within five hours, depending on the sensitivity of the micro-organisms as well as the origin and concentration of the propolis.

The composition of propolis is very complicated and its therapeutic value is dependent upon the origin, cleanliness, storage and the procedure used in the processing of tinctures, ointments, etc. The components of propolis that are alcohol soluble are the biologically effective substances. Seen from the chemical point of view, these active agents are partly Flavone and Flavonoids, which are well-known in medicine for their vessel healing and anti-inflammatory properties, as well as Galangin, Quercetin and Ferula acid and Caffeic acid. Propolis has been demonstrated to be most effective against erysipelas, typhoid and numerous intestinal germs.

Most beekeepers are not happy when a bee colony propolizes frames and hive parts. It certainly makes working the colonies more difficult when the bees stick all moveable parts together. But every beekeeper should weigh the advantages and disadvantages of having bee colonies which are strong propolizers. It is better to make the most of this circumstance and harvest the propolis. Equally advisable is cleaning empty beehives and harvesting the propolis rather than throwing it away.

Some beekeepers burn propolis in their smokers because it smells aromatic and spicy, but it is too precious for that use. The qualitatively purest propolis can be collected only in the summer, after the nectar season. During a bee season it is possible to harvest approximately 50-100g (1¾-3½ oz) of propolis per colony.

There are reputable reports of colonies of heavily propolizing bees, in favourable climates, producing up to 400g (14oz) of propolis per colony. That reflects the results of optimizing operational improvements.

DURING EQUIPMENT CLEANING   3.2

Some beekeepers collect propolis as part of an annual clean-up, to be used subsequently in the making of various tonics for their own personal use. Propolis collected from different hive locations contains differing concentrations of wax. However, this can be separated from the Propolis through cautious melting at approx. 176°F (80°C).

MACICKA and RACKOVA from the Czech Republic made comparative analysis of five Carniolan bee colonies, seeking out the possible annual quantities of propolis crops. Within fixed-time intervals, the bars of Hoffmann frames were scraped clean of propolis. On average, 67g (2.36 oz.) of propolis could be harvested annually per colony using nine frames in the brood chamber and eight frames in the honey super. There is a significant difference between the amount collected from brood boxes and honey chambers. The tests proved 30% more propolis could be harvested in the brood chamber than in the honey super. Three harvests during the year clearly displayed the peak was in June. With only 16-23% wax percentage, the Propolis on the sidebars of the frame is the purest. The top bars contain a propolis which includes 32-56% wax.

SYSTEMATIC PROPOLIS PRODUCTION   3.3

It was the ingenious discovery of Langstroth in the 19th century that made a targeted and systematic propolis collection possible. He found that the bees do not fill spaces of 4.7-9.5mm (3/8-3/16 of an inch) with wax or propolis. For Langstroth, this discovery was significant for the optimal application of his magazine hive management approach. He knew of no practical value concerning propolis production. His hive construction was developed on the basis of preventing the bees from building up the space between the lower and the upper chambers. But, the distance of about 5mm (1/5th inch) is optimal for propolis production. Other authors (GURESOAIE U. MILOIU, 1987) proceed from the assumption that the bees will fill available space to a maximum 3.5mm (1/8th of an inch) with propolis.

The current interest in the therapeutic usage of propolis in both human and veterinary medicine is strengthening both commercial production and marketing.

There are different techniques and methods that can be used for propolis production. Which method you chose depends on the effectiveness, economy and hive-type you prefer. Below you will find some methods that produce satisfying results:

I. **SMALL BOARDS**

One can use these boards in side-opening hives and also in multiple-storey hives. The bees fill small cracks between the boards with propolis. The resulting crop is a relatively pure propolis, which can be harvested easily and several times per year.

II. **BURLAP**

A large piece of burlap can be placed over the top bars of the hive. The bees readily propolize this material, making it possible to harvest twice a year. To remove propolis from the cloth, freeze it until brittle enough to break.

III. **LEIKART FRAME**

Another method to produce propolis was invented by the beekeeper Leikart. He created a frame built from many single hardwood slats with 3-4mm (1/8th of an inch) space between them. Each colony of bees gets such a frame.

When the colony is strong and all other conditions are favorable, two to three propolis collecting frames can be used. This frame can be modified to be used in a multiple storey hive environment, also. The frame is placed on the top of the hive instead of the inner cover, and the inner cover will be placed on the Leikart-frame.

Fig 8   Where beekeepers work with inner cover, noticeable quantities of propolis can
be harvested annually. (Photo Nowottnick).

Fig 9   Where beekeepers work with inner cover, noticeable quantities of propolis can
be harvested annually. (Photo Nowottnick).

Fig 10 At the contact sites of the Hoffmann-frames the propolis is nearly pure. (Photo Nowottnick).

Fig 11 Propolis on the side bars of boards for high hive bottoms. (Photo Nowottnick).

Fig 13   At the top of side-opening hives the collection of propolis is very simple. The beekeeper should place the boards at a distance of 3mm apart from each other, which the bees will propolize. (Photo Nowottnick).

Fig 14   At the top of side-opening hives the collection of propolis is very simple. The beekeeper should place the boards at a distance of 3mm apart from each other, which the bees will propolize. (Photo Nowottnick).

Fig 15   One Leikart frame, Above that the inner cover will be placed.
(Photo Nowottnick).

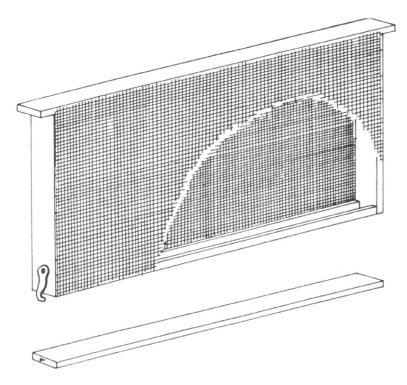

Fig 16   Cut propolis collecting frame with detachable bar on the bottom.
(Photo Nowottnick).

Fig 17   Propolis collecting frame with approx 1.5mm mesh. The moveable bar is removed. (Photo Nowottnick).

Fig 18   A frame with removed under bar. It is very easy to shake out the propolis that has fallen in the inside area. (Photo Nowottnick).

Fig 19   A frame with wire mesh near the hive entrance of a side-opening hive.
(Photo Nowottnick).

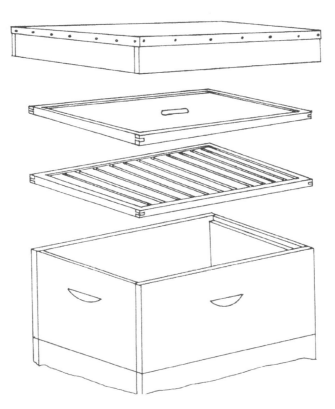

Fig 20   A leikart frame is placed on the uppermost magazine below the inner cover.
(Drawing Nowottnick).

Fig 21 A ventilating wire mesh frame on the top chamber below the inner cover is very useful for propolis collecting. (Photo Nowottnick).

Fig 22   A ventilating wire mesh frame in a freezer. (Photo Nowottnick).

Fig 23   Removing frozen propolis by knocking with the hand and scraping off
with a knife. (Photo Nowottnick).

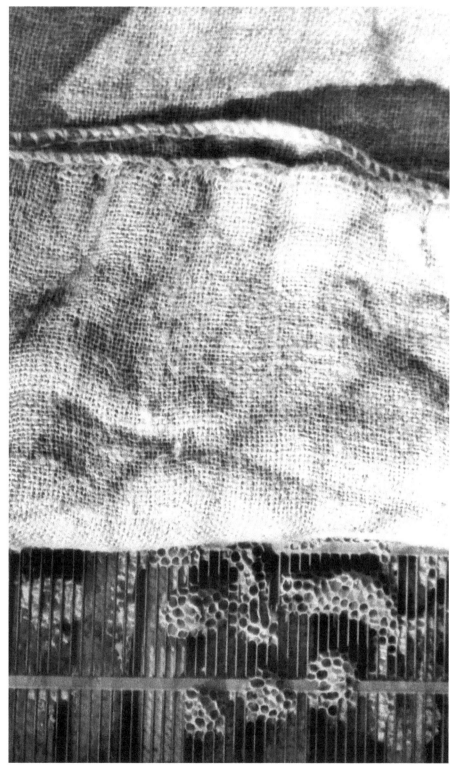

Fig 24   On the queen excluder, placed on top of the last hive body, lie a linen; like this clean sugar sack. (Photo Nowottnick)

Fig 25 A propolized sugar sack on a queen excluder. (Photo Nowottnick)

Fig 26 After removing the sugar sack from the bee colony it will be placed in a freezer. (Photo Nowottnick)

Fig 27   It is very easy to remove the frozen propolis from the sugar sack by kneading the sack. (Photo Nowottnick)

Fig 28   A plastic screen for propolis collecting on top of the upper hive chamber. (Photo Nowottnick)

Fig 29   Rolling down a partly propolized plastic screen from the hive. (Photo Nowottnick)

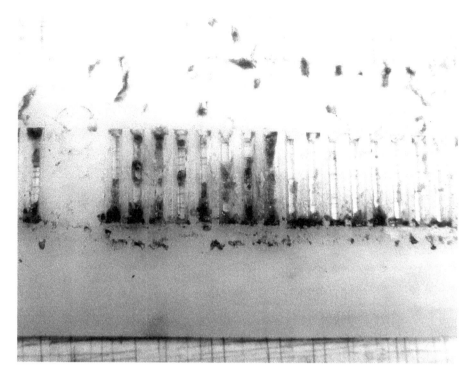

Fig 30   Partly propolized plastic screen.. (Photo Nowottnick)

Fig 31   A plastic screen in a freezer. (Photo Nowottnick)

Fig 32   Harvesting of propolis by rolling the frozen material. (Photo Nowottnick)

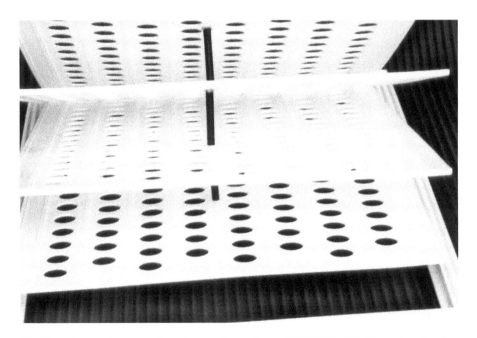

Fig 33   A complete propolis collecting item of type PROPOLMATS (Photo: Les Gera)

Fig 34 One propolized plate of the collector (PROPOLMATS). (Photo Les Gera)

Fig 35 PROPOLMATS covered by a sugar sack or linen. (Photo: Les Gera)

IV. **WIRE MESH FRAME**

A standard sized collecting frame can easily be made. The lower part will need to be detachable. Stability is provided by two thin bars, spanning the width of the frame, on both sides. Wire mesh is stretched over both sides of the frame. This style, as well as the Leikart frames, is placed near the hive entrance of side opening hives.

If the bees have propolized the mesh on both sides, the frame can be placed in the freezer. Once cold, give it a tap or two and the brittle substance will break away in large pieces. Those on the inside area of the frame can be taken out after removing the bottom bar.

V. **VENTILATING WIRE MESH FRAME**

During standard hive management, a wire mesh frame can be placed on the top of the hive, under the inner cover, just as the burlap which was explained above. The size of the mesh should be a maximum of 3mm (1/10th inch). This mesh frame can be used for moving bee colonies too.

VI. **QUEEN EXCLUDER**

OROS from Romania puts a queen excluder on the top of the hive and covers it with linen. During honey harvest, the linen is taken off, chilled, and rubbed to break the propolis loose. Between 100–300g (3½-10½ oz.) of propolis is an expected annual crop using this simple method. A similar method is to lay a bar frame approximately 5mm (1/5th inch) above the top bars of the combs with a mesh cloth or plastic mesh; this will produce similar results.

VII. **PLASTIC SCREEN**

Some producers and sellers of beekeeping equipment offer plastic screens which have slit-shaped openings. They are especially well suited for propolis production in standard hives. The sheet is used with the wide side on the top of the uppermost hive chamber. To harvest, simply twist the frame and the propolis pieces fall out from the slots.

VIII. **IMPREGNATED TEXTILE WEB**

A series of tests was conducted by GURESOAIE &. MILOIU (1987) to determine which material works best for propolis production and where is the best place in the hive to harvest pure propolis. Here are the results of this analysis:

a) Perforated plastic foil with 2.8-3mm (1/10th inch) openings
b) Perforated plastic foil with a base made from cotton
c) Linen coated with plastic material with 1mm weave
d) A thick cotton/linen textile coated with plastic with 1 mm weave

The best material was found to be the plasticized linen. They harvested nine times more propolis using that.

Additionally their study revealed that:
- Most Propolis is stored above the frames.
- Additional ventilation in the hives will stimulate a marked increase in propolis production.
- Avoid removing all propolis from the hive because the bee colony will lose its natural defense mechanism making it susceptible to illness.

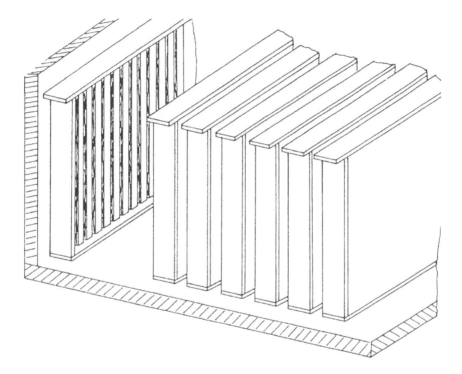

Fig 12   A Cross cut hive with the frames crosswise to the entrance. The first frame in the hive near the entrance is the Lei kart-frame. (Drawing Nowottnick)

IX. MULTIPLE SHEET COLLECTORS

A new propolis collector was introduced in 1996 from New Zealand (Les Gera, Bees-R-Us Ltd., P.O. Box 12194, Hamilton, New Zealand). This tool is constructed of plastic and consists of a frame with four or five perforated plastic sheets. The frame is placed between the top brood box and the super. After one to two weeks, the frame collector is removed or replaced with a new one. Removing the propolis from the sheets is very easy; simply freeze the whole sheet and then break it off.

1. PROPOLIS IN THE PRACTICE OF APITHERAPY   4.1

There has never been discovered a panacea, a cure-all wonder drug, nor do we expect there ever will be. Therefore do not consider propolis a miracle cure. Even with today's knowledge and research, anybody who makes such an allegation not only is a dreamer, but also acts irresponsibly. All of the serious articles and books about the curative effect of propolis state that propolis should be used for the treatment of severe disease only after consulting with and under the recommendation of medical doctors. External application for minor injuries is recommended as standard first aid and doesn't require a doctor's consultation. Never forget, though, that each oral or rectal use of propolis is accompanied by the risk of an allergic reaction. This may be considered a small risk, but not impossible.

Propolis is comparable with traditional antibiotics like penicillin, and it has the power to strengthen the human immune system. Experiments were conducted to ascertain compatibility where test persons were administered high doses. No toxic reactions appeared.

The investigation of the medicinal properties of propolis was triggered by the realization that an incredibly confined social community, as is a bee colony, which maintained maximum cleanliness and normal conditions, had no illnesses appear.

The observations of P. LAVIE support this with the results of his intense research, which displayed that propolis, on approximately 30 microbe stems, is bactericidal. In 1909, N. ALEXANDROW (Russia) published a small article entitled "PROPOLIS AS MEDICINE", which instructed the readers about the treatment of corns with propolis, a method he had known since 1893. The mode of operation was described in the chapter "Propolis in History".

During World War II, propolis was used in two hospitals of Swerdlowsk to promote the healing process of wounds.

- K.LUND AAGAARD (DENMARK) has published a report of the results of four test series, which involved 16,000 people from all over Scandinavia. The results show that 97% of the treated cases were successfully helped, 3% had negative results and three cases produced allergic reactions which led to the discontinuance of the treatment.

- AAGAARD evaluated the effects of propolis treatments and categorized them to the healing of the following complaints: colon inflammation, eye inflammation, infection of the urinary tract, sore throats, gout, rheumatic fever, sinus inflammations, open wounds, colds, flu, stomach cold, bronchitises, periodontosis, earaches, headaches, intestinal infections, mycosis, eczemas, deflection, pneumonia, stomach ulcers, arthritis, lung ailments, stomach virus, flatulence, migraine, gallstones, kidney stones, sclerosis, disturbance of the blood circulation, warts, conjunctiva, chilblains and more.

- LEIPUS (Russia) has written a report based on the information provided by doctors in his homeland, who were using propolis for successfully treating adults and children for such ailments as catarrhs (inflammations) of the respiratory tracts, flu, bronchitis, and bronchial asthma. There has been much success with the healing of surgical wounds generally; abdominal and gynaecological surgery in particular. With propolis, not only the outer purulent wound can be cured, but also injuries of the internal organs because it destroys numerous injurious body toxins. An alcohol based propolis solution made with 20-30% propolis, thinned with warm water was to be taken two to three times daily an hour before meals for the treatment of internal injuries.

Complaints of the respiratory tracts of children were treated with this tincture twice daily before lunch and dinner. A three to five minute long inhalation treatment of propolis in the morning led to equally effective results. The treatment often showed a very quick response. Treatments should be repeated as often as possible without interruption until recovery. According to the memo of I.T. PERSCHAKOW, propolis possesses a strong local analgesic effect, which is (after PROKOPOWITSCH) 3.5 times more effective than the known anaesthesia effects of cocaine and 5.2 times more effective than Novocain.

The application of propolis in the form of alcohol solutions and ointments in stomatologic practise, for example during tooth pulling, was very successful. Ulcers, mycosis and mucous membrane inflammations of the mouth were cured with a propolis solution of 2-4% propolis and ointment with propolis.

- ERSCHAKOW mentioned the healing of different skin diseases like boils, carbuncles and hydroadenitis with an ointment containing 20% propolis, based in a medicinal petroleum jelly (vaseline) and lanolin. A 1% concentration propolis ointment in butter worked well in healing nipple pain of breast-feeding women. An ointment, existing of 10-15% propolis, butter, olive oil or liquid vaseline is effective for the treatment of mucous membrane illnesses of the cervix and other gynaecological ailments. Increasingly positive results were obtained by using a 20-30% alcohol infusion for the inhalation treatment of inflammation of the upper respiratory tracts, including bronchitis, chronic and acute inflammations of the middle ear and chronic tonsillitis. A 30-40% alcohol propolis infusion in a mixture with oil (preferably olive or corn oil) in the ratio of 1:4 is used for the treatment of defective hearing. Always shake before application until a homogeneous fluid develops; an oil-alcohol-propolis emulsion of bright brown color and pleasant smell.

- PERSCHAKOW used cotton balls with an oil-alcohol propolis emulsion in the auditory canal during treatment. For children which are older than 5 years, the cotton balls can be used at night for 10-12 hours and the treatment should be repeated 10-14 times. For adults, this treatment should be used for 2 days. Keep the cotton balls in the ears for 36-38 hours and repeat the treatment 10-12 times. From 381 people who suffered a chronic and acute inflammation of the middle ear, age defective hearing etc., 314 patients showed improvement after treatment. However, PERSCHAKOW does not recommend this kind treatment for adenoids and granulations of the eardrum cave. He points out specifically that the treatment may be considered only after consulting a medical doctor.

- KALMAN (ISRAEL) used propolis for the healing of acute throat complaints. He outlines that he successfully resolved these complaints within a very short time, and additionally, cured a person that had been suffering over ten years by using propolis to close his nasal sinus passages. The periodic use of propolis now grants him free breathing through the nose.

He was able to help an old man who suffered an acute inflammation of the prostate gland using raw propolis. The patient was to chew propolis for ten minutes and then swallow it. After the treatment, the man did not require surgery.

In 1979, Kalman's wife became seriously ill. The doctors informed him she had cancer of the liver and pancreas that was so advanced that she could live only another three weeks to three months. The tumor was a very aggressive form and treatment using chemotherapy and cobalt irradiation was not possible because of other illnesses (diabetes and angina pectoris). After consulting with his daughter, also a medical doctor, the wife was started on a propolis treatment. The patient received 1-2g of propolis a day. A medical examination made a year later showed that the tumour had not grown. Her pain was reduced and she has lived many years longer.

A six-year old boy, whose white blood count had fallen drastically and was no longer responding to medicine, started receiving daily dosages of propolis for a month. The result of this treatment was a normalization of the white blood corpuscles.

Initially, propolis was predominantly used for respiratory diseases (inflammations) and the treatment of nose and ear problems. Current practice shows all virus infections, including the Asiatic flu, can be effectively treated with this bee product. Propolis application is effective on bad cases of idiopathic-thrombosis as well as acute sinusitis (nasal sinus inflammation). Sinusitis was cured within a month by the daily dosage of 1.5g of propolis.

- KALMAN ascertains, in agreement with doctors, that propolis could stop the growth of cancer and could even eliminate the cancer tumour totally, when used together with a chemotherapeutic treatment. Even some cases of leukaemia have been cured with propolis.

- POSDZIECH outlines the effective treatment of a severe, continuing, seven week long, agonizing case of hiccups. He dispensed his patient 10 drops of a 15% tincture of propolis. Immediately after taking the drops with some fluid, the hiccups were banished. The duration of the effect corresponds with the number of drops dispensed. If one gives the patient in one dose 1g of propolis, it will remain effective approximately 18 hours. The treatment is repeated as long as needed, until the hiccups are eliminated completely. If one takes granules instead of the drops, the effect lasts for only an hour.

- MR. SUCHY, S. SCHELLER and I. ZAWADSKI report on their investigations into the possible applications of propolis in gynaecology. The conclusions show that propolis extracts accelerate the healing processes of wounds after gynaecological operations that would otherwise be difficult to cure. Propolis extract shows good results for the treatment of inflammations of the vagina and the cervix through pathogen (pathogenic) fungi or from bacterial infections. The best results from the application of propolis extracts were achieved within the first 7-10 days. In rare cases an extension of the therapy caused allergic reactions.

From 1964 to 1972 L.N. DANILOW and his staff members explored the curative effect of propolis on certain skin diseases. Propolis was used as an ointment and as tincture as part of a study made on 680 persons with different skin ailments. The ointment was thinly applied to the area of the wound and then bandaged. As a rule, the bandage was changed daily. In addition, 30-40 drops of propolis tincture were dispensed a half hour before meals daily. The treated patients had chronic exanthema and limited neurodermitis. Each affliction was localized on the back and heel of the hands as well as the bend in the knees and arms. The treatment extended over a month but after only 5-6 treatments, the itchiness eased and the skin became soft and more elastic. Simultaneously the patients reported that they felt strengthened and had improved sleep and appetite. The result of the treatment was positive in 90.1% of the cases.

- W. WASSILEFF, ST. MANOWA-KANAZIREWA, W. TODOROFF and DRJANOWSKI explored the efficacy of propolis on Intertrigo (*Dermatitis intertriginosa*). The treatment, which was applied to babies between the ages of one to three months, showed good results. This disease exhibits symptoms of irritation of the skin in the folds of the buttocks and femorals. The skin reddens, erythema and secretions appear and as a consequence, secondary infections with pustules appear. This disease is accompanied by restless sleep, nervousness and crying. Propolis was used in a 30% ointment which was applied directly on the affected areas twice daily for two to six days.

A 55% propolis solution in an aerosol was used for the treatment of tropical ulcers of the lower limbs in the surgical ward of Hospital No. 22 in Kiev (Ukrainia). Twenty five patients participated in this treatment; they were all cured. On the right lower leg of a 70-year-old female patient, the skin had died and developed an ulcer. The traditional medicinical treatment was ineffective. After treating with inhalations and placing a bandage with propolis ointment on it, the ulcer was healed.

Wash the surface of surgical sites and other wounds with hydrogen peroxide, pat dry with some clean gauze and loosely cover for 12 hours with propolis ointment on a bandage. This supports granulation and promotes healing of the wound.

The publication of new scientific results enriches and validates propolis application in human medicine, which in some countries is still based partly on following time-honoured traditions.

Multifarious are the recipes and processes. There is no doubt that propolis brings relief and healing.

APPLICATIONS   4.2

Propolis is excellently qualified for use as a prophylactic treatment as well as a curative. Propolis assists, intensifies and initiates the immune system. Following are a few methods of using propolis that are low risk, if not harmless. If an allergic reaction occurs, the treatment must be stopped immediately.

I.   **2.1. PAIN KILLING REMEDY**

Havsteen identified propolis as a natural aspirin. Propolis has the power to relieve pain because it contains flavonoids, similar to those contained in synthetic pain remedies, which inhibit the development of prostaglandin, which is what causes pain. Propolis has the advantage that its Flavonoides are of natural origin and do not have any side effects on the digestion and nervous system, unlike synthetic analgesics. Do not disregard that prostaglandins possess an important function, because pain is a sign that something is wrong.

II.   **HERPES AND FLU TREATMENT**

These illnesses are caused by viruses which can be treated successfully with propolis. It is recommended to take propolis mixed into honey as a prophylactic, or preventative treatment. Also herpes blisters can be cured successfully by covering with a propolis solution.

III.   **THROAT AND PHARYNX TREATMENT**

In case of complaints in the throat one can use a tincture which is produced as follows: 25-30g (1oz) cleaned propolis is mixed with 80ml (3oz) of 96% alcohol. This mixture has to be shaken vigorously until the propolis has dissolved, then filtered through a paper filter into a bottle to remove the smaller dirt particles and the non-solvable beeswax parts. This filtrate should be bottled in dark glass jars and stored in a cool, dry place. For treating sore throats and colds, mix six to seven drops of the tincture into a glass of lukewarm water and gargle with this solution three to four times per day.

A very effective and simple treatment for the flu, particularly when used at the first signs of the illness, is to mix a teaspoon of propolis tincture into a glass of chamomile or lime-tree blossom tea and add a little honey or lemon-juice. Use this three to five times a day.

For internal application, one can chew propolis in it natural state. Alternately, cleaned propolis powder can be kneaded and formed into a "sausage." Place this into the freezer. When frozen, the sausage can be grated into small pieces. For inflammations in the mouth and throat, chew a teaspoon of propolis and for stomach-intestine-illnesses chew a teaspoon several times a day [KEMPF].

IV.   **WOUND TREATMENT**

Propolis ointment or tincture used on slow and poorly healing wounds will promote quick healing. The wound needs to be covered three to four times per day with propolis tincture.

V.   **ARTERIOSCLEROSIS TREATMENT**

This illness, which represents a thickening and narrowing of the blood vessels, can be deadly. Cholesterol, a major contributor to this condition, is too available and over consumed. Poor nutrition and excessive use of alcohol and tobacco are also prime contributors, which presents a dreadful clinical picture for our culture. For testing purposes, a group of patients were given propolis capsules daily over the period of a month. At the end of a month, the cholesterol levels had dropped (HAVSTEEN, ROHWEDDER, 1987).

VI.   **WOMEN'S CONCERNS**

One of the first medicinal uses of propolis was for regulation of menstruation and relief of cramping. Taking propolis tincture regularly eases and regulates in a natural way. The treatment of vaginal inflammations using propolis has been successful. SCHELLER used it as suppositories. The effects of the treatment appeared after about two weeks.

VII. **JOINT COMPLAINTS**

An inflammation of the tendons, tendon sheaths and muscles occurs from overworking the joints. Rub the joints with propolis ointment daily to ease and eliminate these complaints. Indeed, during the first days of treatment the pain may increase, but that fades away usually around the third day or so. Resting the damaged area will aid the recovery.

VIII. **HEMORRHOIDS TREATMENT**

This suffering can be very painful and often ends with an unavoidable operation. Propolis ointment will at least have a palliative effect and in many cases can also provide healing. The ointment should not be used just externally, but should be applied as far as possible internally. The micro-organisms present internally are eliminated by the active substances of the propolis. As a matter of course, attention is to be paid through regular and thorough cleansing. According to [KEMPF] the use of an alcoholic propolis tincture with 10% of propolis for the external application is very effective. Dissolve 100g (3½oz) propolis powder in 630g (22oz) pure alcohol and filter through a coffee filter. The solution can be stored, without loss of effectiveness, in a dark bottle in a dark place. This tincture can be applied to fresh and infected wounds. When mixed with castor oil at a ratio of 1:1, it can be used on haemorrhoides, where there are skin cracks, fissures, and varicosities.

IX. **PROSTATE**

This illness first occurs primarily when men are in their 40s. The complaints, accompanied by inflammations, in many cases can be soothed or cured with propolis.

X. **RHEUMATIC COMPLAINTS**

HAVSTEEN reports on the testing of a group of patients who all had problems with their spinal columns, either in the neck or chest. One half of the group was treated with a chemical compound and the other one with propolis ointment. Three times more patients treated with propolis showed a clear improvement within the first week.

Arthritis of the hip and knee joints is very painful. During an investigation in Munich, about 80% of all arthritis patients treated with propolis ointment were pain-free within a week.

XI. **MENOPAUSAL SYMPTOMS**

The age-related diminishing function of the ovaries results in a regression of estrogen. This appearance is designated as Menopause and is characterized by heart complaints, "hot flashes" and depression. In a Polish hospital, propolis was used to stimulate the metabolism. The affects of this treatment were unexpectedly good.

XII. **LOCAL ANAESTHETIZATIA**

According to TZAKOFF (1975) propolis has a stronger effect in quelling localized pain than cocaine or novocain. Both are very customarily used medicinally.

XIII. **IMMUNE SYSTEM**

Propolis increases the production of antibodies and strengthens their influence on the immune system. Even though there are no definitive confirmations by science, the preventive use of propolis would provide relief and stabilization of the body's own defence system (EXNER, 1990). KIWALKINA and BUDARKOWA (1969, 1975) said propolis is a non-specific, generic stimulant of immunological reactions.

XIV. **DECONTAMINATION**

According to EXNER (1980), there is discussion about the use of propolis helping to decontaminate and neutralize the human body of such toxins as noxious emissions, environmental chemicals, pesticides and heavy metals, for example lead, nickel, platinum, cadmium and mercury.

XV. **ANTI-CANCER EFFECT**

Specific derivates of caffeic acid are an important component of propolis, primarily caffeic acid phenyl ester, shortly CAPE, (Koenig, 1988.) In laboratory experiments with healthy human cell cultures, there was no cytotoxic effect associated with this substance. But, in contrast, tumor cells were attacked (GRUNBERGER, 1988). Further numerous tumour cell lines were tested for sensitivity to CAPE (Koenig, 1988).

XVI. **FOR TISSUE REGENERATION**

The flavonoids in propolis greatly improve circulation as well as acting as powerful anti-inflammatory agents. Pharmacological studies show that propolis has a supporting affect on tissue regeneration, for example: bone tissues, dental enamel and the keloid.

XVII. **THE ANTI-OXIDIZING CHARACTERISTICS OF PROPOLIS**

According to USCHKALOWA and MURICHNITSCH (1995), because of the presence of flavonoids, propolis is able to hamper, and even prevent the fermentation process and formations of mould. That is more and more important for both the food and cosmetics industry as well as for medicine and biology. Propolis is being used for preserving specific foods, such as deep-frozen pizza, in some countries (Japan).

XVIII. **USE WITH OTHER ANTIBIOTICS**

Investigations have shown that the effectiveness of some antibiotics is intensified by the additional application of propolis (CHERNYAK, 1973; KIWALKINA, 1969; KIWALKINA and GORSHUNOVA, 1973). The inhibitorical (inflammation hampering) effect of biomycin, tetracycline, neomycin, polymyxin, penicillin and streptococcus aureus was about 10-100 times higher after the addition of propolis into the growth medium. Similar synergistic effects were observed after the use of mixtures of propolis and streptomycin, penicillin and furagin against types of staphylococcus (CHERNYAK, 1973; KIWALKINA and GORSHUNOVA, 1973).

ALLERGIES   4.3

The application of propolis can lead to side effects and allergic reactions. These reactions can differ depending on the origins of propolis, the plant species visited and the season. Most people have no problem, but before using propolis one should be tested for compatibility. Allergic reactions can manifest themselves as increased skin sensitivity as well as with acute sickness. When prescribing propolis as a treatment, great importance must be attributed to the allergic anamnesis (case history of the illness). Only a medical doctor can initiate treatments with propolis. Some authors sight that one of 1000 will suffer an allergic reaction, and others report the numbers are closer to one of 2000.

Animal experiments have shown that Propolis does not possess carcinogenic substances.

RECIPES   5.1

Raw propolis can have many contaminates such as wax parts, wood splinters, bee parts, etc. Fresh raw propolis should be cleaned before using. Following are some procedures to accomplish that.

A) The simplest method involves shaking up small pieces of propolis in a bowl. The conspicuous pollutants will surface and can be picked out. Repeat as necessary. Another means is simply to spread it out on a pad and the pollutants can be removed from there.

B) IANNUZZI (1995), recommended placing the raw propolis in a pot, fill it up with cold water and stir with a fork. The parts floating on the surface are skimmed off (try a tea-strainer). Then the contents are poured through a sieve and rinsed with water again. This should be done until nothing more floats on the surface; only the propolis and water remain. This is poured

Fig 36   Propolis deposited on a Leikart frame, ready to be harvested.
(Photo Nowottnick)

Fig 37   Purified propolis ready for making tincture. (Photo Nowottnick)

Fig 38   Using a paper funnel, a glass is filled with purified crude propolis.
(Photo Nowottnick)

Fig 39   Propolis is mixed with 96% medical alcohol from the pharmacy.
(Photo Nowottnick).

Fig 40   After filled, the bottle need to be sealed well with a cork and aluminium foil.
(Photo Nowottnick).

Fig 41   Shake vigorously once a day. (Photo Nowottnick)

Fig 42   After the propolis has dissolved in the alcohol, the crude tincture is strained
through a paper filter. (Photo Nowottnick)

Fig 43   After filtering, the tincture can be filled into small bottles.
(Photo Nowottnick)

Fig 44   A small bottle with pipette. (Photo Nowottnick)

through a fine sieve and let the water drain from the propolis. Then spread it on a piece of absorbent material where it will dry within 24 or 48 hours. When done, the pollutants left, such as small wax crumbs and such, can be taken out with tweezers.

C) Propolis harvested from the frames, inner covers and other hive parts has a high share of wax and other pollutants. Place in a pot with water and warm to 185°F (85°C). The wax will melt and float to the surface while the propolis sinks. By stirring repeatedly, all imbedded wax will separate from the propolis. Remove from the heat and once cooled, the wax layer can be removed. The water is poured out leaving the propolis on the bottom of the pot. It will be easier to remove the propolis if put into a freezer for awhile (WHITE, 1993). Cleaning by using heat may lessen the quality of the Propolis.

TINCTURES OF PROPOLIS   5.2

Tincture of propolis is used predominately for oral health. It is used in both concentrated and diluted forms for treating a variety of complaints as well as prophylactically. Propolis is also administered externally to combat various symptoms of illnesses, like the small blisters of herpes. The following suggestions are just the beginning of the possibilities for use of propolis. (Normally propolis tinctures of 5% to 10% are used.)

I. **BRAILEANU, GHEORHIU, POPESCU, VELESCU (RUMANIA)**
    HAVE PUBLISHED THE FOLLOWING RECIPE:
    Ingredients:
    10g (1/3rd oz.) propolis - pulverized and sieved
    100ml (3 1/3rd oz.) alcohol - 30-95%

The propolis and the extracting fluid (alcohol) are poured into brown vessels and then sealed tightly and kept at room temperature for 10 days. During this time they are shaken two to three times daily. When dissolved, the alcohol solution is strained through a paper filter. The tinctures are stored in dark brown flasks for seven days at room temperature and then filtered once more. The bacteriological researchers pointed out the tinctures made with alcohol of 70-90% are the most effective at dissolving the propolis.

II. Purified and crushed propolis is placed in a vessel and filled with twice the amount of 96% alcohol. Cover and leave standing for 10-14 days, shaking daily. At the end of this time, either strain the tincture through a paper filter or use as is. Before each use the contents of the bottle should be shaken well. Use a sterile cotton swab or pad to apply.

III. KULHANEK (1988) recommends mixing 70ml (2 1/3rd oz.) pure ethanol with 30g (1 oz.) finely grated propolis in a brown bottle. The mixture is shaken daily during a four to five week curing period, after which it is strained through a coffee filter. Take 15-20 drops daily, after a meal. This is used to strengthen the resistance powers of the body and to prevent flatulence, etc.

IV. HELMING-JACOBY (1991) suggests using crude propolis with 90% alcohol that just covers the propolis. For 100g (3-1/5th oz.) propolis you will need 120ml (4oz.) alcohol. Leave the mixture stand for about eight days, shaking daily. The tincture is then left undisturbed for some weeks. During that time, the undissolved elements will settle to the bottom and the tincture can be decanted, carefully. It is quicker to use a paper filter but this method produces a tincture with a higher concentration.

V. GROHMANN (1988) makes his tincture of propolis the following way: 100-200g (3½-7 oz.) crude propolis is dissolved in 1 litre (34oz.) of alcohol. Regular stirring or shaking accelerates the dissolution. Subsequently the mixture is strained through a paper or cotton wool filter.

VI. ZANGERL (1990) suggests using frozen propolis. Fill an electric coffee mill and grind it to powder. The propolis powder is placed in a jar and mixed with three times as much 96% medical alcohol. This mixture remains stored for 14 days in a warm room. It must be shaken once or twice a day after which it is strained through a cloth or a paper filter and decanted into small bottles.

Fig 45   The deep-frozen propolis can be pulverised by a coffee mill. (Photo Nowottnick)

Fig 46   Look in the coffee mill, filled with frozen propolis. (Photo Nowottnick)

Fig 47   Within a few seconds the propolis turns to powder. (Photo Nowottnick)

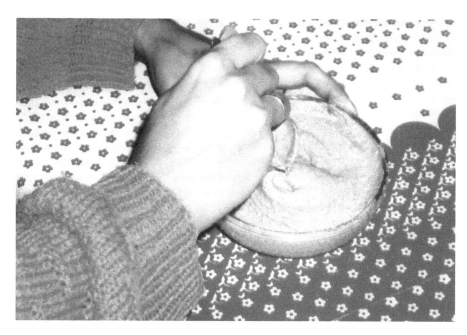

Fig 48   The powdery propolis is stirred into the ointment base and mixed. (Photo Nowottnick)

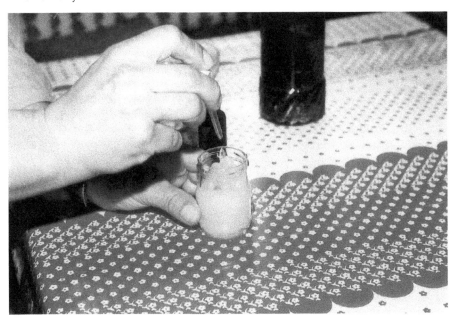

Fig 49   Adding propolis extract into the ointment base by pipette. (Photo Nowottnick)

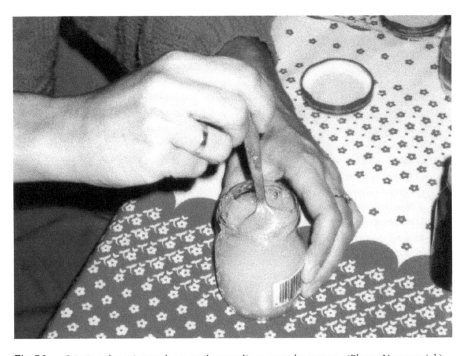

Fig 50   Stirring the mixture base and propolis extract by spoon. (Photo Nowottnick)

a) For sore throats and complaints of the upper respiratory tracts, one to two teaspoons of tincture gargled long enough for it to be swallowed with the saliva. It is better to use the tincture thinned with water, because of the sensitive mucous membranes of the mouth. The application should be used at least three times daily.

b) Colds can be cured with a three times daily application of 15-20 drops of the tincture of propolis. The tincture is mixed with warm water and drank. In addition, gargling with this solution speeds the healing process.

c) One should coat lingering and badly curing wounds three to four times daily with the tincture, which will promote healing.

d) It is reported that stomach ulcers are also curable with tincture of propolis. One mixes 30-40 drops into 100-125ml (3-1/3rd -4oz.) warm milk. This should be consumed twice a day, an hour before a meal.

e) Liver and kidney inflammations can allegedly be treated with good success by drinking 30-40 drops in warm tea twice a day. There are still contradictory opinions about this because the resins of the propolis as well as the alcohol would not be very favourable for liver and kidneys.

f) Arthritis should be treatable through the following cure: On the first day one takes one litre (34oz.) pure and unsweetened apple juice mixed with one tablespoon of honey and 6ml (1/5th oz.) tincture of propolis. On the second day, drink one litre (34oz.) of grape juice mixed with 1 tablespoon of honey and 6ml (1/5th oz.) tincture of propolis. On the third day, 10 teaspoons fruit vinegar is mixed into a glass of nettle tea. These are taken three times daily, respectively.

VII. DANILOW made a tincture from 100g (3½ oz.) propolis and 500ml (17oz.) 96% pure medical alcohol. Both were poured into a glass bottle and stored for 14 days in a dark place, where it was shaken up now and then. After 10 days it should be filtered through a gauze cloth.

VIII.   The propolis tincture made by GEORGE TAMAS is as follows; 20g (2/3rd oz.) of propolis combined with 100ml (3½ oz.) 96% medical alcohol. The mixture is poured into dark bottles and stored for about 15 days in a cool, dark place. Later the mixture has to be filtered. For treating the flu and colds, one measures 20 drops of propolis tincture into a glass filled with lukewarm water.

IX. For making a simple propolis tincture, mix 100-200g (3½-7oz.) raw propolis and one litre (34oz.) medical alcohol 96%. The solution should be filtered through a coffee filter.

PROPOLIS EXTRACTS   5.3

It is not useful to use tinctures of propolis for the production of ointment. The amount of tincture required to be effective when mixed into an ointment would be impractical. Therefore, as much propolis as possible needs to be dissolved in as little alcohol as possible to be of use in an ointment. Following are two methods which are relatively simple.

I.   **TWO STEP PROCEDURE OF GROHMANN:**

To get a higher concentrated propolis solution, a larger quantity of raw propolis is required. Higher concentrated propolis solutions made with very little alcohol are better for the production of ointments. Fresh propolis is added to the already filtered propolis tincture. This highly concentrated tincture is difficult to strain through a coffee filter. Using a vacuum pump helps accelerate the filtration, however.

II. **CONCENTRATING BY EVAPORATION**

The simplest way to produce high percent propolis extract for making ointments is to let thicken in a natural way, like as is practiced by HELMING-JACOBY. In this process, tinctures of propolis can be thickened up to $2/5^{th}$ of the earlier volume. One lets the tincture of propolis stay in an open vessel at a warm place for some weeks. The final product is a syrupy concentrate with about 90% propolis.

OINTMENTS WITH PROPOLIS   5.4

For the production of ointments, it is important that the propolis is distributed throughout the ointment homogeneously. For this reason, the use of propolis powder is not effective. Therefore propolis extract is the best component.

I.   A mixture of cold-pressed olive oil and beeswax in a proportion of 4:1 works well as an ointment base. The beeswax is melted carefully in a water bath, mixed with olive oil and warmed carefully until the mixture is clear. Then remove from the heat and stir constantly until the mixture has stiffened like an ointment. Later, a measured and weighed amount of concentrated propolis tincture is added drop by drop and stirred for some minutes. The completed ointment can be poured into salve or balm containers and stored in a cool, dry place. The propolis tincture used should be of 35% from which is made an ointment with 5% propolis. Using this method, one needs 5g ($1/5^{th}$ oz.)propolis tincture for 100g ($3\frac{1}{2}$ oz.) of ointment.

Here a mathematical example of GROHMANN:

The plan is to produce 1000g (35oz.) of ointment containing 5% propolis. Used is a propolis solution of 35%. For 1000g (35oz) of ointment one needs 0.05 x 1000g (35oz.) = 50g (1 2/3rd oz.) propolis. But in 1g of 35% propolis solution are only 0.35% contained. Therefore from this propolis tincture one needs 50: 0.35 = 142.85g (5oz.). To make 1000g (35oz.) of 5% propolis ointment one needs 142.86g (5oz.) of 35% propolis tincture.

II. In 100g (3½ oz.) of ointment base is added 10g (1/3rd oz.) of concentrated propolis tincture. This kind of ointment can be spread on the skin easily. At the same time, it is possible to buy ready-made ointments containing biological raw materials in cosmetics stores or drugstores. Such ointments can be used for making propolis ointments, in which case only the concentrated propolis tincture is added.

III. In a water bath are melted 240g (8½ oz.) of lanolin and 150g (5oz.) beeswax (from virgin wax combs). To this add 520g (18oz.) warmed cold-pressed sunflower or olive oil. Cool the mixture by stirring constantly and then add 90g (3oz.) propolis extract.

IV. HELMING-JACOBY made a small change to the recipe above to make a more pliant ointment. Required are 200g (7oz.) lanolin, 100g (3 ½ oz.) beeswax, 80g (3oz.) lanolin alcohol and 530g (18½ oz.) oil.

V. KULHANEK makes a mixture of 100g (3½ oz.) vaseline and 10g (1/3rd oz.) lanolin. Both are stirred constantly in a 68°F (20°C) warm water bath and 20-25ml (2/3rd –4/5th oz.) 20% alcohol tincture of propolis is added. By constant stirring and further warming, the alcohol will be evaporated from the homogenized ointment.

VI. One takes 20g (2/3rd oz.) ointment base, made from 10g (1/3rd oz.) lanolin and 10g (1/3rd oz.) vaseline, and adds 10g (1/3rd oz.) propolis extract. Remove from the hot water bath and let the ointment reach a semisolid state. After homogenizing this, add the remaining 70g (2½ oz.) of vaseline. If the propolis extract has been stirred well, the ointment will have a mustard color and an aromatic smell.

VII. Another quite simple recipe for producing a propolis ointment is: make a tincture by covering propolis with 96% alcohol and shake well until you have a satisfying solution. This then is added drop by drop to the ointment base (for example vaseline). If you want the ointment firmer, you can add some beeswax while making your ointment base in a warm water bath.

VIII.   ZANGERL uses 100g (3½ oz.) white vaseline, 100g (3½ oz.) water-free lanolin, 50g (1 2/3rd oz.) larch turpentine and 20 drops dwarf pine oil to make his ointment base. These ingredients are mixed together, warmed to a maximum of 104°F (40°C) and liquefied. Then 25g (1oz.) of propolis powder and 10ml (1/3rd oz.) propolis extract (viscous 1:1) are added and stirred. This ointment is pored in liquid form into small, dark containers.

IX. ZANGERL recommends the making of another propolis ointment.
He takes 40g (1 2/5th oz.) of purified and very finely chopped propolis (propolis extract would be even better), 60g (2 1/10th oz.) ointment base made with 30g (1oz.) lanolin, 30g (1oz.) vaseline and 30g (1oz.) beef suet. The addition of 10 drops of dwarf pine oil and 15 drops of larch turpentine improves it's effectiveness at opening the pores and ultimately improves circulation.

X. POCINKOVA's extraction ointments can be made with different ointment bases; for example vaseline, vaseline-lanolin in proportion of 10:2, sunflower oil and others. The ointment base is warmed and small pieces of propolis added. The amount of propolis depends on the concentration desired. Warm this mixture in a water bath, stirring every 20-30 minutes. When the mixture has cooled slightly, filter it. With this method, not all of the active agents of the propolis are extracted and they are partly destroyed by the application of heat. Therefore, this form of the ointment production isn't recommended.

XI. Propolis ointment with salicylic acid (POCINKOVA)
100g (3½ oz.) crushed or pulverized propolis is combined with 20-30ml (2/3rd –1 oz.) pure medical alcohol of 95%. Repeated stirring will dissolve the propolis within two to three days. Then 500g (17½ oz.) vaseline is warmed to the boiling-point, cooled down to approx. 140°-158°F (60°-70°C) and the propolis alcohol mixture added. The mass is stirred about 15 minutes on low heat until the alcohol has evaporated. Then screen and filter the mixture while warm. This product obtains 25g (4/5th oz.) salicylic acid during stirring.

XII. Propolis ointment (after WHITE)
Ingredients:
15g (1/2 oz.) of beeswax
60g (2 oz.) of olive oil
15g (1/2 oz.) propolis powder
15g (1/2 oz.) of honey
The ingredients are placed in a pot, which is placed in a water bath, and melted and then stirred until it has cooled.

XIII.   Propolis ointment (after GROHMANN)
Ingredients:
200g (7 oz.) of pure beeswax
800g (28 oz.) of olive oil
170g (6 oz.) 30% propolis tincture

The ointment base is produced as follows: The wax is melted in a water bath without overheating it. The liquefied wax is then mixed with the olive oil. If this mixture is still opaque, then reheat it in warm water bath until it is clear. After this, let it cool down and then stir to give it a homogeneous consistency. When the cream has cooled down to 86°F (30°C), add the propolis solution, stirring constantly. Leave it uncovered for a couple of days to allow the alcohol to evaporate before filling your individual jars. The propolis has to be dissolved in pure, 96% alcohol to create the tincture. The non-soluble components can be filtered out by pouring through filter paper. The propolis tincture should contain 30g (1oz.) of propolis in 100ml (3½ oz.). The finished product will contain approx. 5% propolis.

Olive oil has an advantage that it is readily absorbed into the skin. But some people don't like olive oil or don't like the smell. Almond oil also can be used as an alternative. The beeswax in the ointment doesn't penetrate into the skin, but it forms a protective layer on the skin. The consistency of this ointment can be changed by simply adding more or less beeswax.

PROPOLIS POWDER   5.5

HELMING-JACOBY writes about making a powder using a very fine propolis dust, which can be made by grinding in a coffee grinder. Use frozen propolis, to minimize the stickiness. The propolis dust is sifted through a fine meshed sieve. The fine propolis powder is especially well suited for making powder using a basic ratio of 1:4. Use the following mixture: 20g (3/4th oz.) propolis powder, 44g (1½ oz.) talcum, 20g (3/4th oz.) Bolus Alba (white clay earth), 10g (1/3rd oz.) Terra silica (pebble earth) and 6g (1/5th oz.) zincoxide.

HONEY WITH PROPOLIS   5.6

A Russian doctor wrote a recommendation for honey in the Polish bee journal of "Pszczelarstwo" (11/1995), where five, 10, 15 or 20% propolis was added. The five, 10, 15 or 20g of propolis must be melted in a ceramic or stainless steel pot in a water bath until it is the consistency of glue. Honey is added to make 100g

(3½ oz). This means, there are 5g propolis added to 95g of honey, 10g of propolis to 90g honey, and so on. These components are stirred over a water bath until a homogeneous looking mixture has formed. After, the propolis-honey should be strained and stored in a dark, air-tight glass jar and kept cool. Honey with propolis smells like propolis and strengthens the body's defenses against illnesses. It will heal wounds and alleviate pains, as well as impede the development of bacteria, viruses and fungi's. This propolis infused honey should be taken three times a day, a half hour before meals. If using the 5% honey, take one teaspoon three times a day, if using the 10-15% solution, take ½ teaspoon, and if using the 20% solution, take ¼ teaspoon, three times a day. Heavier people can increase the dosage. Usually one should take the propolis honey for a period of between five and 30 days. With tuberculosis, administer for six to eight weeks and if necessary, repeat after a two-week break. The propolis honey should be used until full recovery if treating pneumonia, lungs catarrh, tonsillitis or larynx catarrh.

1.6 FACE MASKS   5.7

According to a recipe by C. KROCHMAL, the antibacterial effect of propolis is effective against greasy skin.

Ingredients:
60ml (2oz.) buttermilk
5g (1/5th oz.) propolis
50g (1/5th oz.) of honey
10ml (1/3rd oz.) lemon juice

The buttermilk and the finely crushed propolis are mixed and left to sit for an hour. Place the mixture into a little cooking pot and warm over low heat until the propolis has dissolved. Remove from the heat and add the remaining ingredients. After the mixture has cooled, it is applied with a cotton ball on the skin. The mask is left on until it feels dry, then washed with warm water. The above measurements can be adjusted according to need.

HAIR CARE REMEDIES   5.8

For oily hair (C. KROCHMAL) propolis helps to clear away excessive grease of the scalp and hair.

INGREDIENTS:
60g (2oz.) of ordinary yohgurt
10g (1/3rd oz.) propolis
125g (4 ½ oz.) of honey
1 egg yolk

Mix yogurt and propolis powder and let sit for an hour. Then place into a small pot and warm on low heat. Let boil for a couple of minutes until the propolis has dissolved. Remove from the heat and add the remaining ingredients. Stir until cooled. The mixture is applied to damp hair, left on for 30 minutes, and then washed out with a mild shampoo.

LIPSTICKS WITH PROPOLIS   5.9

Ingredients:
15g (1/2 oz.) of beeswax
5g (1/5th oz.) propolis powder
5g (1/5th oz.) of lanolin
40g (1½ oz.) of olive oil
A few drops Wintergreen

Mix all the ingredients, except the propolis, and heat while stirring in a water bath over low heat. Then remove from the heat, add the propolis powder and stir until cooled.

---

2. PROPOLIS IN VETERINARIAN MEDICINE

GINDL reports in the journal "Bee Father" in 1994 that he could heal fever, diarrhoea, cough and lack of appetite in young pigs and piglets with honey and propolis. The veterinarian recommends this over the use of conventional antibiotics which are very expensive and take longer to be affective.

OWCZARCZYK (1995) writes of a report in the Polish bee newspaper that propolis can effectively cure fungal diseases in bee colonies.

One mixes a soup spoon of 10% propolis tincture into a litre (34 oz.) of sugar solution 1:1 (one part sugar to one part water.) This sugar-propolis mixture is administered to the bee colonies three days in a row. After a week, no trace of the fungal infection will be visible. The author has administered such treatments in the Fall and Spring for several years. However, avoid using this treatment during a honey flow so that no sugar gets into the stored honey.

SUMMARY

Stress is expected in our vision of modern life, even if it impares our pursuing a healthy lifestyle. Propolis contains the ability to counter and protect us from the harmful effects of stress. It is incumbent upon beekeepers to spread the word about the wonderful effects of propolis. It is important to consult your family doctor first. It is of inestimable value to win the doctor's interest and partnership. The conclusions and efforts presented here are this author's attempts to expand the knowledge of the beneficial characteristics of propolis. The picture I offer is founded on results; free from mystification and superstition.

Still, research continues. My fervent wish is the same as the wishes of many beekeepers, that propolis research would spread and become legitimized in still more countries. As propolis usage increases and becomes accepted, beekeepers will be able to enjoy an economic improvement. The more we can promote propolis usage, the more we can afford to keep bees, as well as benefit the health and well-being of the public, Even though propolis has not yet gained wide-spread support in many countries, we trust in a better future.

BIBLIOGRAPHY

AAGAARD, L.K. (1975) DER NATURSTOFF DER PROPOLIS-QUELLE DER GESUNDHEIT (DIE PROPOLIS) *APIMONDIA-publishing house BUCHAREST*

ARTOMASOWA, A.W. (1975) ALLERGIEN BEI PROPOLIS (DIE PROPOLIS) *APIMONDIA-publishing house Bucharest, Romania*

AUTORENKOLLEKTIV (1985) DER SCHWEIZERISCHE BIENENVATER *Sauerländer publishing house, Aarau, Switzerland*

AUTORENKOLLEKTIV (1985) GRUNDWISSEN FÜR IMKER *VEB DEUTSCHER LANDWIRTSCHAFTSVERLAG BERLIN*

AUTORENKOLLEKTIV (1980) APITHERAPIE HEUTE *APIMONDIA-publishing house Bucharest*

BOLSCHAKOWA, W.F. (1976) ZU DEN ALLERGENEN EIGENSCHAFTEN DER PROPOLIS (NEUES IN DER APITHERAPIE) *APIMONDIA-publishing house Bucharest*

BOLSCHAKOWA, W.F. (1975) PROPOLISANWENDUNG IN DER DERMATOLOGIE (DIE PROPOLIS) *APIMONDIA-publishing house Bucharest*

BRAILEANU, E.; GHEORGHIU, A; POPESCU, A.; VELESCU, GH. (1975) FORSCHUNGEN ÜBER BESTIMMTE PRÄPARATE MIT PROPOLIS (DIE PROPOLIS) *APIMONDIA-publishing house Bucharest*

CAILLAS, A. (1975) PROPOLIS (DIE PROPOLIS) *APIMONDIA-publishing house Bucharest*

CIZMARIK, J.; CIZMARIKOVA, R.; MANTEL, J. (1975) PRÄPARATE MIT PROPOLIS (DIE PROPOLIS) *APIMONDIA-publishing house Bucharest*

CIZMARIK, J.; MACICKA, R.; MANTEL, J. (1975) ANALYSE UND KRITIK DER THEORIEN ÜBER DIE PROPOLISBILDUNG (DIE PROPOLIS) *APIMONDIA-publishing house Bucharest*

DANILOW, L.N. (1975) BEHANDLUNG BESTIMMTER HAUTKRANKHEITEN MIT PROPOLIS (DIE PROPOLIS) *APIMONDIA -publishing house Bucharest*

DROEGE, G. (1984) DAS IMKERBUCH *VEB DEUTSCHER LANDWIRTSCHAFTSVERLAG BERLIN*

EXNER, J. (1990) BLÜTENPOLLEN UND PROPOLIS-IHRE QUALITÄT UND ANWENDUNG TEIL II *BIENENWELT 32(7): 176-181*

EXNER, J. (1990) BLÜTENPOLLEN UND PROPOLIS-IHRE QUALITÄT UND ANWENDUNG TEIL III *BIENENWELT 32(8/9): 200-204*

EXNER, J. (1990) BLÜTENPOLLEN UND PROPOLIS-IHRE QUALITÄT UND ANWENDUNG TEIL IV *BIENENWELT 32(10): 226-229*

EXNER, J. (1990) BLÜTENPOLLEN UND PROPOLIS IN FRAGE UND ANTWORT *BIENENWELT 32(11): 238-239*

GRIMM, G. (1977) EIN TROPFEN NEKTAR *VEB DEUTSCHER LANDWIRTSCHAFTSVERLAG BERLIN*

GROHMANN, F. (1988) DIE VERARBEITUNG VON PROPOLIS ZU CREME UND TINKTUR *BIENENWELT 30(6): 169-172*

GURESOAIE, J.; MILOIU, J. (1991) UNTERSUCHUNGEN ZUR GEWINNUNG VON PROPOLIS *NEUE BIENENZEITUNG 2(9): 36-37*

HAUSEN, B.M.; WOLLENWEBER, E. (1988) PROPOLIS ALLERGY *CONTACT DERMATITIS No.19: 296-303*

HELMING-JACOBY, L. (1991) PROPOLISZUBEREITUNGEN *DEUTSCHES IMKER-JOURNAL 2(6): 228-231*

IBRA RESEARCH NEWS No. 9 -PROPOLIS- *IBRA, HILL HOUSE BUCKS*

Ikeno, K.; Ikeno, T.; Miyazawa, C. (1991) EFFECTS OF PRPOLIS ON DENTAL CARIES RATS *CARIES RES.No.25: 347-351*

JOIRISCH, N.P. (1978) DIE WELT DER BIENEN *ECON VERLAG WIEN UND DÜSSELDORF*

KALMAN, CH. KOPIEN VON MANUSKRIPTEN, DIE ZU APIMONDIA KONGRESSEN UND SYMPOSIEN DURCH DEN VERFASSER VERLESEN WURDEN 1.SOME FIELD EXPERIENCES WITH PROPOLIS 2.APITHERAPY SUCCESS IN ISRAEL

KIWALKINA, W.P. (1976) BILANZ UND AUSSICHTEN DER PROPOLISFORSCHUNG (NEUES IN DER APITHERAPIE) *APIMONDIA VERLAG BUKAREST*

KULHANEK, V. (1988) PROPOLIS-NATUREHEILMITTEL AUS DEM BIENENVOLK *SCHWEIZERISCHE BIENENZEITUNG 111(2): 57-72*

LAVIE, P. (1975) DAS ANTIBIOTIKUM DER PROPOLIS (DIE PROPOLIS) *APIMONDIA VERLAG BUKAREST*

LEIPUS, J.K. (1975) PROPOLIS,EIN WIRKSAMES HEILMITTEL (DIE PROPOLIS) *APIMONDIA VERLAG BUKAREST*

MACHACKOVA, J. (1988) THE INCIDENCE OF ALLERGY TO PROPOLIS IN 605 CONSECUTIVE PATIENTS PATCH TESTED IN PRAGUE *CONTACT DERMATITIS No.18: 210-212*

MACICKA, M.; Zlatica; Rackova (1983) PROPOLISQUANTITÄT UND –QUALITÄT ABHÄNGIG VOM SAMMELORT *APIMONDIA VERLAG BUKAREST*

MAKASCHWILLI, Z.A. (1975) PROPOLIS IN DER GESCHICHTE (DIE PROPOLIS) *APIMONDIA VERLAG BUKAREST*

MIHAILESCU, N.N. (1976) APITHERAPIE BEI PROSTATALEIDEN (NEUES IN DER APITHERAPIE) *APIMONDIA VERLAG BUKAREST*

MIZEAHI, A. (1982) BIOLOGICAL PROPERTIES OF PROPOLIS-AN OVERVIEW *PROCEEDINGS OF THE NORTH AMERICAN APITHERAPY SOCIETY, LINTHICUM MARYLAND*

MORSE, G.D. (1975) ÜBER PROPOLIS UND IHRE VERWENDUNG IM BIENENVOLK (DIE PROPOLIS) *APIMONDIA VERLAG BUKAREST*

MRAZ, CH. (1976) APITHERAPIE IN DEN USA (NEUES IN DER APITHERAPIE) *APIMONDIA VERLAG BUKAREST* Neumann, D.; Götze, G.; Binus, W. (1986) KLINISCHE STUDIE ZUR UNTERSUCHUNG DER PLAQUEGINGIVITISHEMMUNG DURCH

PROPOLIS *STOMATOLOGIE DER DDR 36(12): 677-681*

OGREN, W. (1986) DON`T THROW OUT THAT PROPOLIS *SPEEDY BEE Nr. 1/1986*

OTTOTSKI, L. (1975) DIE SPURENELEMENTE DER IMKEREIERZEUGNISSE (DIE PROPOLIS) *APIMONDIA VERLAG BUKAREST*

PERSCHAKOW, I.T. (1975) BEHANDLUNG DER SCHWERHÖRIGKEIT MIT PROPOLIS (DIE PROPOLIS) *APIMONDIA VERLAG BUKAREST*

PESTSCHANSKI, A.N. (1975) BEHANDLUNG EINIGER LEIDEN MIT HILFE EINER PROPOLISLÖSUNG (DIE PROPOLIS) *APIMONDIA VERLAG BUKAREST*

PESTSCHANSKI, A.N. (1975) KONZENTRIERTE PROPOLIS (DIE PROPOLIS) *APIMONDIA VERLAG BUKAREST*

POPRAWKO, S.A. (1976) CHEMISCHE ZUSAMMENSETZUNG DER PROPOLIS, IHRE HERKUNFT UND STANDARDISIERUNGSFRAGEN (NEUES IN DER APITHERAPIE) *APIMONDIA VERLAG BUKAREST*

POPRAWKO, S.A. (1975) CHEMISCHE ZUSAMMENSETZUNG DER PROPOLIS (DIE PROPOLIS) *APIMONDIA VERLAG BUKAREST*

POSDZIECH, H. (1985) EIN NEUER VERWENDUNGSZWECK FÜR PROPOLIS *DIE BIENE 10/1985*

PRISITSCH, W.N. (1975) EIN NEUES DESODORANT (DIE PROPOLIS) *APIMONDIA VERLAG BUKAREST*

ROHWEDDER, D.; HAVSTEEN, B.H. (1987) PROPOLIS - DER STOFF AUS DEM GESUNDHEIT IST

SUCHY, H.; SCHNELLER, S. (1975) ERGEBNISSE DER PROPOLISANWENDUNG IN DER GYNÄKOLOGIE (DIE PROPOLIS) *APIMONDIA VERLAG BUKAREST*

TROSHEV, K.; KOZAREV, I.; MARKOV, D. (1990) REGULATION OF TRAUMATIC SKIN WOUND HEALING BY INFLUENCING THE PROCESS IN THE NEIGHBOURING ZONE OF THE WOUND *ACTA CHIRURGIAE PLASTICAE 32(3): 162 - 172*

TURELL, M.J. (1975) PROPOLIS - EIN ARZNEIMITTEL DER ZUKUNFT? (DIE PROPOLIS) *APIMONDIA VERLAG BUKAREST*

VASILCA, A.; MILCU, E. (1976) LOKALE BEHANDLUNG CHRONISCHER GESCHWÜRE MIT PROPOLISEXTRAKTEN (NEUES IN DER APITHERAPIE) *APIMONDIA VERLAG BUKAREST*

WALLNER, W. (1986) IMKER PRAXIS *ÖSTERREICHISCHER AGRARVERLAG WIEN*

WASILEFF, W.; MANOWA-KANAZIREWA, ST.; TODOROFF, W.; DRJANOWSKI, ST. (1975) PROPOLISBEHANDLUNG BEI INTERTRIGO UND MONILIASIS BEI SÄUGLINGEN (DIE PROPOLIS) *APIMONDIA VERLAG BUKAREST*

WEISS, K. (1981) DER WOCHENENDIMKER *EHRENWIRTH VERLAG MÜNCHEN*

ZANGERL, A. (1986) PROPOLIS UND SEINE HEILWIRKUNG *SELBSTVERLAG FLIRSCH*

ZANGERL, A. (1990) PROPOLIS UND IHRE HEILWIRKUNG *SELBSTVERLAG FLIRSCH*

Lightning Source UK Ltd.
Milton Keynes UK
UKHW031827230219
337901UK00004B/31/P